Jordana Souza Paula Riss
Josimar B. Ferreira

Influence of homeopathic medicines and preparations on lettuce

Jordana Souza Paula Riss
Josimar B. Ferreira

Influence of homeopathic medicines and preparations on lettuce

Seedling development

Imprint
Any brand names and product names mentioned in this book are subject to trademark, brand or patent protection and are trademarks or registered trademarks of their respective holders. The use of brand names, product names, common names, trade names, product descriptions etc. even without a particular marking in this work is in no way to be construed to mean that such names may be regarded as unrestricted in respect of trademark and brand protection legislation and could thus be used by anyone.

Cover image: www.ingimage.com

This book is a translation from the original published under ISBN 978-613-9-61736-4.

Publisher:
Sciencia Scripts
is a trademark of
Dodo Books Indian Ocean Ltd. and OmniScriptum S.R.L publishing group

120 High Road, East Finchley, London, N2 9ED, United Kingdom
Str. Armeneasca 28/1, office 1, Chisinau MD-2012, Republic of Moldova, Europe
Printed at: see last page
ISBN: 978-620-7-69964-3

CONTENTS

1   INTRODUCTION 2
2   THEORETICAL FRAMEWORK 4
3   MATERIAL AND METHODS 15
4   Results and discussion 22
CONCLUSION 34
FINAL CONSIDERATIONS 35
BIBLIOGRAPHICAL REFERENCES 36

1 INTRODUCTION

Homeopathy was created in the 18th century by the German doctor Samuel Hahnnemann. It is a natural therapeutic method that stimulates the body's defense system in order to achieve balance. It is based on four principles: similarity, experimentation on healthy individuals, minimum doses and a single medicine. It aims to treat all living beings in different areas, as it is based on the Law of Similars (HAHNEMANN, 1996; KHUDA-BUKHSH, 2006; COSTA et al. 2009).

Homeopathy applied to agriculture, also known as Agrohomeopathy, has emerged as a promising alternative for the implementation of sustainable crops, as it dispenses with the use of pesticides, contributing to food and environmental safety. Homeopathic therapeutic practice was legalized under Law No. 10831, by the Ministry of Agriculture and Supply through Normative Instruction No. 46 of October 6, 2011, which regulates its application in organic farming (BRASIL, 2011). Today, several research groups in the country, with a greater concentration in the South and Southeast regions, have been conducting experiments with plants and verifying the effects of the homeopathic method on plants (ROLIM, 2009).

According to Bonato (2009), research using homeopathic medicines and preparations on plants has been carried out by farmers and researchers, with promising results in increasing resistance to diseases, tolerance to unsuitable physical conditions, breaking seed dormancy and producing healthy seedlings, covering all sectors of agriculture. However, there are many difficulties faced by these researchers, including the absence of a Homeopathic Materia Medica of Plants (MMHP), a different scenario to human and veterinary medicine, which have specific works (CARNEIRO et al, 2011a).

Homeopathic medicines and preparations are formulated from substances from the three kingdoms: plant, animal and mineral (CARNEIRO et al., 2011a). They are obtained through the process of dilutions followed by succussions (dynamization or potentization).

Brazil currently leads the world in pesticide consumption and the National Cancer Institute warns of food contamination (LONDON, 2011). We are therefore faced with the challenge of producing healthy food, with the least possible impact on the environment, in an economically and socially sustainable way. Andrade (2010) states that the use of homeopathy in vegetable production is seen as a technology aimed at the innovative market, due to its low dependence on external inputs, the fact that it provides healthy food and contributes to environmental safety. Thus, positive results from the action of homeopathy in seedling development could contribute to the consolidation of this science in agriculture, providing producers with a sustainable and safe production alternative.

Seedling production is the main stage in vegetable growing, as the plant's performance depends on it (ROSSI, 2005). In Acre, the volume of lettuce production varies throughout the year, due to climatic conditions, the lack of agronomic research and consequently the incipient development of new technologies to meet the immediate needs of vegetable growers, resulting in a significant drop in production (SILVA, 2010).

The aim of this study was to evaluate the influence of different strengths of the homeopathic medicines *Arnica montana, Calcarea carbonica, Carbo vegetabilis, Silicea terra, Phosphorus, Pulsatilla nigricans* and the homeopathic preparation of açai residues on the development of seedlings of two lettuce cultivars, "Verònica" and "Regina".

2 THEORETICAL FRAMEWORK

2.1 HOMEOPATHY

Homeopathy is a word of Greek origin meaning "similar disease" (*homois=similar*, pathos=disease). Created in 1796 by the German doctor Christin Friedrich Samuel Hahnemann, it is a natural therapeutic method that stimulates the body's defense system in order to achieve balance (HAHNEMANN, 1996; KHUDA-BUKHSH, 2006; COSTA et al. 2009).

Also known as the science of high dilutions, homeopathy is traditionally used on humans, but its application extends to different areas such as agriculture, veterinary medicine and dentistry. It aims to treat all living beings, soil and water and is based on the Law of *Similars* enunciated by Hippocrates in the 4th century BC, "*Similia similibus curantur*" (like cures like) (ALVES, 2014).

According to Bonato et al. (2014), the foundation of homeopathy states that there is no such thing as disease, only illness. Disease in Hahnemann's conception is a disturbance of the vital force (the life that animates the being).

The science of Homeopathy is based on four principles: similarity, experimentation on healthy individuals, minimum doses and a single medicine (HAHNEMANN, 1996; KHUDA-BUKHSH, 2006; COSTA et al. 2009).

2.1.1 Similarity

When translating Dr. Wilian Culen's materia medica, Hahnemann was intrigued by the chapter on the use of quinine (*Cinha officinalis*) in the treatment of malaria (ROSSI; HAHNEMANN, 2009). Hahnemann disagreed with Cullen's explanations and decided to try small doses of the plant and observe its effects. When he took it, he developed the same symptoms as malaria, which disappeared when he stopped using the plant. Based on the aphorism of Hippocrates (460 - 350 BC) "like cures like" (MODOLON, 2010), he came to the conclusion that the substance that produces certain symptoms in the healthy organism is capable of curing the same symptoms in the sick organism.

According to Hahnemann, homeopathy is based on nature's laws of healing. Therefore, when a substance that has the same pattern of imbalance studied in healthy plants is applied to an unbalanced plant, the unbalanced plant will return to its balanced state (Figure 1).

In homeopathy, any disturbance caused to the plant, whether by biotic or abiotic factors, first acts on the plant's vital energy (responsible for homeostasis).

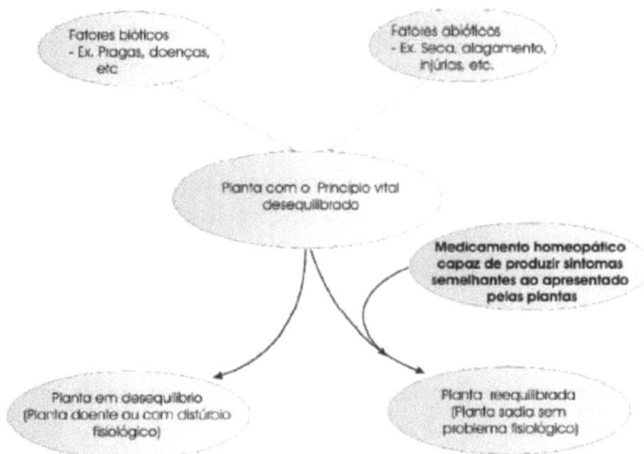

Figure 1. Factors that interfere with the biology of the plant and the action of the homeopathic medicine through the "Law of similars" (Source: BONATO, 2009).

The action of homeopathic medicines in stressful situations is aimed at balance, and plants will respond through self-regulation (the ability to produce effects in the opposite direction to the action) which will stimulate secondary or primary metabolism (CASALI et al., 2006).

2.1.2 Experiments on healthy individuals

After observing the effects of quinine on his body, Hahnemann felt motivated to test other substances in the search for new medicines. He experimented on friends, family and himself. Through his various experiments, the foundations of homeopathy were laid down by his mentor in the book *Organon of the Art of Healing* and other publications (CARNEIRO et al., 2011a).

According to Moraes (2009), during experimentation, increasing diluted doses of the substance under study are introduced into the healthy individual, and the responses are called primary action, causing signs and symptoms, which are noted in order to characterize the substance.

To avoid any kind of interference, the experiments follow the double-blind procedure, i.e. the experimenter and the applicator do not know which treatment is being applied (CARNEIRO et al., 2011a).

2.1.3 Minimum doses

With the initial aim of avoiding intoxication or other aggravating factors that the substances used in the experiments could cause, Hahnemann decided to dilute them. Diluting alone wasn't enough, as the medicinal potential was diminished, so Hahnemann proposed a pharmacotechnical method for preparing homeopathic medicines called dynamization or potentization (dilution and vigorous succussion), in other words, agitation adds kinetic energy to the preparation. The more dynamized the substance, the greater the therapeutic effect and the lower the toxic effect (ROSSI; HAHNEMANN, 2009).

In the pharmacotechnical method of dynamization, by convention, the centesimal scale (Hahnemannian centesimal or CH) is used, in which each dilution is carried out in a ratio of 1:100 (solute:solvent) followed by 100 stirrings (succussions) (ALVES, 2014):

- 1 part matrix substance (mother tincture or TM) + 99 parts inert carrier (solvent) + 100 sucrose = 1CH (10^{-2} mol);

- 1 part of 1CH + 99 parts of inert vehicle + 100 sucrose = 2CH (10^{4} mol); and so on;

- 12CH = 10^{-24} mol of the matrix substance, i.e. no matter.

From 12CH onwards, Avogadro's number (6.023×10^{23}) is exceeded and there is no longer any matter (molecules or ions of the original substance), but only memory (energy).

The memory or energy (drug information) contained in the infinitesimal doses of ultradiluted substances is responsible for promoting changes in organic systems similar to those of the original substance and are explained by the laws of Quantum Biophysics (CARNEIRO et al., 2011a).

2.1.4 Single Medication

According to Vithoulkas (1980), in the experimentation protocol, Hahnemann warns of the importance of trying one homeopathic medicine at a time in order to obtain a reliable report on the signs and symptoms caused by the preparations.

The aim is to arrive at the ideal medicine, known in homeopathy as the *Similimum*, which is the medicine that covers most of the signs and symptoms of the individual being treated (ALVES, 2014).

2.2 AGROHOMEOPATHY

Agrohomeopathy can be defined as the use of the homeopathic method in agriculture (ROSSI; HAHNEMANN, 2009). This same name is also used to refer to the use of homeopathy in the whole of agriculture.

2.2.1 History of Agrohomeopathy in Brazil

Studies into agrohomeopathy are recent, although the first experiments in this area were carried out by Kolisko and collaborators in Germany in 1923, stimulated by the ideas of Rudolf Steiner, but it was only in the 1990s that these experiments became more frequent.

Homeopathy in vegetables was first regulated in Brazil by Normative Instruction No. 007 of May 17, 1999, which legalized its application in organic farming, recommending it for the control of diseases and pests (BRASIL, 1999) and later by Law No. 10.831, the Ministry of Agriculture and Supply's Normative Instruction no. 46 of October 6, 2011 regulated its application in organic farming, recommending it both for the control of diseases and pests and for the physiological rebalancing of plants (BRASIL, 2011).

The Federal University of Viçosa (UFV) in the state of Minas Gerais is the pioneering Brazilian public institution in researching and disseminating experiments using homeopathic methods on plants. This work began in 1998 (CASALI et al., 2011).

The first master's thesis on homeopathy in Brazil, entitled "Homeopathy on the growth and production of coumarin in chambà (*Justicia pectoralis* Jacq)", was defended on December 13, 1999, by agronomist Fernanda Maria Coutinho Andrade, under the supervision of Prof. Dr. Vicente Wagner Dias Casali, at the Phytotechnics Department of the Federal University of Viçosa (ANDRADE, 2000).

In 2003, the first course to train homeopathic agronomists began at the Center for Advanced Studies in Homeopathy (CESAHO) in the city of Piracicaba (ROLIM, 2009).

In 2004, the research and extension activities in agrohomeopathy carried out by UFV were certified by UNESCO/Bank of Brazil Foundation as an effective social technology. Being a social technology means being simple, cheap and accessible to all farmers. Being effective means solving the problem it was designed to solve (CASALI et al., 2011).

According to Casali and collaborators (2011), this certification is due to the fact that homeopathy is an impactful method with proven results that solves the social problem of the rational/ecological use of land to produce healthy food, respecting biodiversity and eliminating pesticides from farms.

In 2004, a project was started at the Agronomic Institute of Paranà (IAPAR) dealing with pathogenetic experimentation on plants, with the aim of starting to draw up the 1st Homeopathic Materia Medica das Plantas (MMHP) (CARNEIRO et al., 2011a).

Also in 2004, the Attorney General's Office determined that homeopathy was not just a medical specialty and legalized the activity of the popular homeopath (CASALI, et al., 2011).

By 2011, UFV had 28 dissertations and theses defended in the field of Homeopathy (SILVA, 2014).

Congresses, conferences, seminars, meetings and extension courses have been held throughout Brazil. In 2015, the Federal University of Acre (UFAC) hosted the 17th Brazilian Seminar on Homeopathy in Organic Agriculture, which took place on the Floresta campus in Cruzeiro do Sul (UFAC, 2014). According to Andrade and Casali (2011), the event has been held since 1999, with the aim of updating knowledge, sharing knowledge, providing information on alternative practices and publicizing homeopathy in organic agriculture. It is promoted by UFV and partners under the coordination of Dr. Vicente Wagner Dias Casali, a professor at UFV and a researcher with the National Council for Scientific and Technological Development (CNPq).

Also in 2015, the topic of homeopathy applied to agriculture was included for the first time in the 70th Congress of the *Liga Medicorum Homeopathica Internacionalis* (LMHI) (verbal information)[1] . The main event was a conference given by agronomist Dr. Carlos Moacir Bonato, as well as oral presentations and posters on the same theme. Until then, the LMHI had only been attended by doctors, dentists, veterinary doctors and pharmacists.

Several research groups in Brazil, with a greater concentration in the South and Southeast regions, have conducted experiments with plants and found the effect of homeopathy on vegetables (ROLIM, 2009). This has resulted in the publication of scientific dissemination materials such as articles and instructional texts, using differentiated communication resources that are accessible to different audiences (CASALI et al., 2011).

2.2.2 Development of homeopathy in agriculture

According to Bonato (2009), research using homeopathy on plants has been carried out by farmers and researchers, with promising results in increasing resistance to diseases, tolerance to unsuitable physical conditions, breaking seed dormancy and producing healthy seedlings, covering all sectors of agriculture. The effects of homeopathy on agriculture go beyond what is recommended by the Normative Instruction, as it is also possible to treat the soil according to Bonato et al. (2014) and water according to Araùjo et al. (2013).

---

1 News provided by Carlos Moacir Bonato, at the 70TH *CONGRESS OF THE LIGA MEDICORUM HOMEOPATHICA INTERNACIONALIS* (LMHI), in Brazil, in August 2015.

However, there are many difficulties faced by researchers who set out to carry out experiments applying homeopathy to plants, because they don't have a Homeopathic Materia Medica of Plants (MMHP). This is a different scenario to human and veterinary medicine, which has specific works (CARNEIRO et al, 2011a).

Carneiro et al. (2011a) point out that experiments with homeopathy must comply with the principles described by Hahnemann. Simply dynamizing a substance and evaluating its effects on plants does not mean that one is working with homeopathy on plants.

An example of work that followed these principles is that of Betti et al. (2003) cited by (FERREIRA, 2011; CARNEIRO et al., 2011b), who used *Arsenicum album* (made from Arsenic) to reduce the severity of tobacco mosaic, caused by the tobacco mosaic virus (TMV). The drug was chosen on the principle of similarity. They tested $_{23}$ (Arsenic Trioxide) in high concentrations on tobacco leaves and the lesions caused by the substance resembled the lesions resulting from the hypersensitivity reaction induced by TMV. The authors observed that the homeopathic treatment of the plants with As O_{23} was satisfactory.

Until there is an MMPH that provides guidance based on the principle of similarity - and based on symptoms that are characteristic of plants - one of the alternatives for researchers when choosing the homeopathic medicine to use on plants is to use analogies between the Materia Medica (symptoms in humans) and plants (BONATO, 2009).

Because of the physiological differences between humans and plants, care must be taken when establishing these analogies, which must be unquestionable in order to give credibility to the therapeutic indications. It is also essential to cite the materia medica that was the source of consultation when establishing the analogy (CARNEIRO et al., 2011a).

An example of experimentation by analogy is that of Andrade (2000) who, when analyzing the history of Chambà (*Justicia pectoris*), a medicinal plant, found that this species showed similarities with the pathogenesis (signs and symptoms) of *Arnica montana*, a medicine indicated for organisms with defensive behavior and supersensitive to touch after traumatic conditions.

Nunes (2005) highlights the importance and possibilities of using and the advantages of experimenting on plants: the great diversity, i.e. you can study everything from perennial plants to short-cycle plants; the simplicity of researching the effect on seeds and seedlings; the possibility of working with larger populations; it allows you to evaluate various medicines (respecting individualization) and various strengths or frequencies, as well as forms of application.

Dynamized preparations are widely used in olive growing. Bonato and Silva (2003) evaluated the effects of *Sulphur* in different potencies on the growth and production of radish. They found that the potencies 5CH, 12CH, 30CH and 1MCH (1:1000 dilution) increased leaf length and plant height; 12CH, 30CH and 1MCH increased root diameter.

CASTRO et al (1999 cited by SANTOS, 2011) used *Phosphorus* on radish and observed an increase in the fresh mass of the aerial and root parts. Luis and Moreno (2007) studied the effects of homeopathic medicines called Calcarea in 30CH dynamics on the vegetative growth of spring onions and found that *Calcarea fluorica* 30CH provided a 45% increase in the fresh weight of the vegetable compared to the control.

It is worth mentioning that research using new homeopathic preparations (substances that are not included in the Materia Medica) made from local resources has gained recognition, as they are seen as a strategy for sustainability. Casali et al. (2011) state that farmers who use homeopathic methods in the field are very creative, born experimenters and excellent observers of the dynamic processes of nature, and are important elements in the formulation of new knowledge, technologies and proposals.

Homeopathic farmers often use homeopathic preparations made from pest insects or parts of the diseased plant, known as biotherapics[2] , because they have easy access to this material (ANDRADE et al., 2012). They act according to the Law of Equals, i.e. equals are balanced by equals, and are accepted in homeopathy as long as they are manipulated according to homeopathic pharmacotechnics (CARNEIRO et al., 2011a).

Other studies have also shown the efficiency of this technique for controlling various pests, such as the one carried out by Fazolin et al. (2002) when they applied a dynamized preparation made from the bean leafhopper (*Certoma tingomarius*), a defoliating insect and the main pest of the bean crop in the state of Acre. They observed that the insect did not prefer to feed on bean plants (*Phaseolus vulgaris* L. cultivar Carioquinha) that contained this treatment.

And Andrade et al. (2012) aimed to evaluate the response to dynamizations of the Justicia preparation on the growth and production of coumarin in *Justicia pectoralis*. They found that the fresh matter of leaves and stems, the total fresh matter and the yield of coumarin

---

2 The term bioterapico - which in Greek means *bios:* life, living being (animal or vegetable) and *therapeia:* treatment - replaces the terms nosode and isoterapic. They are medicinal preparations obtained from biological products, chemically undefined (secretions, excretions, tissues and organs, pathological or not, products of microbial origin) and allergens (CARNEIRO et al., 2011a).

varied according to the strengths. These and other studies demonstrate the effects of homeopathic preparations on plants, confirming that homeopathy on plants, even under the Law of Equality, has satisfactory effects.

The use of medicines or homeopathic preparations on plants stimulates the defense system of organisms so that they resist diseases, insect pests, the impacts of climatic or environmental factors and also act to vary the synthesis of active ingredients, change energy patterns and improve production. It promotes balance without extinguishing: viruses, fungi, bacteria, insects and other types of agents (OLIVEIRA et al., 2012).

Andrade (2010) states that the use of homeopathy in vegetable production is seen as a technology aimed at the innovative market, due to its low dependence on industrialized inputs, the fact that it provides healthy food and contributes to food and environmental safety.

Agrohomeopathy is a science that has a lot to be developed and researched, so it is necessary to have an interactive approach between researchers from different fields - biological, veterinary, pharmaceutical and agronomic (ROSSI et al., 2004).

Homeopathy research applied to plants can still reveal an infinite number of compounds, plants, animals (parts, secretions), minerals and microorganisms, with the possibility of homeopathic application. The search for new preparations can help in the search for efficient and ecologically coherent strategies for homeopathic crops.

2.2.3 Homeopathic medicines and preparations

Mother tincture (MT) or Matrix are the names given to the precursor solution of homeopathic medicines and preparations.

Homeopathic medicine or homeopathy is any pharmaceutical presentation intended to be administered according to the principle of similarity, for preventive and therapeutic purposes, obtained by the method of dilutions followed by successive succussions and/or triturations (FARMACOPEIA HOMEOPATICA BRASILEIRA, 1997).

Dynamized substances not described in the Homeopathic Materia Medica are classified as "homeopathic preparations" according to the Brazilian Association of Homeopathic Pharmacists - ABFH (2007).

Homeopathic medicines are obtained from the plant kingdom, the animal kingdom and the mineral kingdom (CARNEIRO et al, 2011a). Examples: Vegetable- *Arnica montana, Belladonna, Pulsatilla nigricans, Chamomilla*; Animal- *Apis melifica* (bee), *Cantharis vesiatoria* (beetle), *Lachesis muta* (Surucucu snake venom); and Mineral- *Arsenium album, Aurum metallicum, Cuprum metallicum, Phosphorus, Sulphur* (BONATO et al., 2014). Of the medicines used in this work, we can highlight the following properties:

• *Arnica montana*: is of plant origin, obtained from whole plants; recommended for reducing the stress of pruning, pruning, grafting, hail, cold winds, excessive heat, transplants and other injuries (BONATO et al., 2014; REZENDE, 2009; BOFF, 2008; ANDRADE, 2007).

• *Carbovegetabilis*: of plant origin, obtained from plant parts

(wood). Recommended for the slow recovery or death of plants after transplanting, mechanical injuries and rot; aids in the greater absorption of nutrients and reduces the incidence of diseases; useful medicine for making plants stronger and more robust (CARNEIRO et al., 2011a; BONATO et al., 2014).

• *Phosphorus*: of chemical origin, obtained from inorganic substances. It is indicated for pest attacks, reduced photosynthetic rate, necrosis, leaf congestion and weak plants (BONATO et al., 2014). It influences the development of new tissues and the production of shoots and leaves (CARNEIRO et al., 2011a).

• *Silicea terra* : of chemical origin, obtained from inorganic substances. It is recommended as a general tonic and transplant shock; for weakened plants, slow growth, stunted growth (CARNEIRO et al., 2011a). Strengthening the cell wall, reducing diseases and pests (BONATO et al., 2014).

• *Calcarea carbonica* : of animal origin, obtained from the inner part of the shell of *Ostrea edulis* L. It is indicated for seedlings that are sensitive to cold; delay in the emission of new roots, plants that are slow to grow or flower (CARNEIRO et al., 2011a). Reduces dependence on limestone in crops and improves the absorption and use of calcium by plants (BONATO R, obtained from whole plants). Recommended for boosting productivity and reducing flower abortion (BONATO et al., 2014).

• Açai waste: *Euterpe precatoria* is a species of native açai found in the state of Acre, popularly known as single açai. It belongs to the Arecacea family and the *Euterpe* genus. The fruit of this palm has great nutritional and energy value. The waste resulting from pulping represents 83% of the fruit and is made up of seeds and shells (Teixeira et al., 2004). This pulping is done in appropriate machines, with the addition of water. According to Teixeira et al. (2004) the decomposition of the seeds results in good quality organic fertilizer.

2.3 ALFACE

Lettuce is a crop traditionally grown by small producers, which gives it great socio-economic importance. It is one of the most consumed vegetables in Brazil (HENZ; SUINAGA, 2009). According to Rossi (2005), the production of seedlings is an important stage in the cultivation of vegetables, as the plant's performance depends on it; a poorly formed seedling gives rise to a plant with limited production.

2.3.1 Choosing cultivars

In Acre, the volume of lettuce production varies throughout the year due to climatic conditions, the lack of agronomic research and the incipient development of new technologies to meet the immediate needs of producers, resulting in a significant drop in production (SILVA, 2010).

According to Lédo (2000), the 'Regina' and 'Verônica' cultivars are promising for cultivation in the state of Acre during the dry season, but during the rainy season they have low yields and quality.

With the intention of bringing a significant economic improvement to Acre's farmers, allowing their agricultural activity to be profitable all year round and, as a result, sustainable. Sustainable because homeopathies leave no solid residues and do not harm the environment, animals, humans or the plants themselves. Profitable, because homeopathic formulations require small quantities of raw materials, resulting in minimal use of natural resources and low costs. As an example, 30 milliliters of a homeopathic medicine costs R$ 20.00 reais (budget carried out in homeopathic pharmacies in Rio Branco - AC, on 11/14) and produces a high yield - approximately 1mL of the medicine is used for 1L of water (CASALI et al., 2014).

2.3.2 Homeopathic studies with lettuce

In the production of lettuce seedlings, Rossi et al. (2003) evaluated the effects of three frequencies (24, 48, 72 hours) of application of the Carbo 30CH medicine and found that in the 48-hour interval there was a 22% increase in the dry weight of the leaves compared to the control. These same authors also checked the effects of Carbo in different strengths on lettuce seedlings grown in two different environments, and found that the 100CH strength increased the dry mass of the aerial part of plants grown in a shaded environment and that 6CH and 200CH increased the height of plants grown in the sun (ROSSI et al., 2006).

In the development of lettuce, Grisa et al. (2007a) found that the 6CH strength of the *Arnica montana* medicine increased the weight of the fresh matter of the aerial part, and the 6CH and 12CH strengths increased the weight of the dry matter of the aerial part, with regard to the number of leaves and plant height, the results were not significant.

The use of homeopathic preparations on the growth and productivity of lettuce was reported by Josè and Cuèllar (2009), who studied the influence of dynamized MB-4 rock flour. According to the authors, MB-4 rock flour is a natural product derived from nutrient-rich rocks. Although they didn't obtain statistically significant results, it was possible to notice a tendency towards superiority in the results of green and dry mass of the aerial part when compared to the control when applying the 12CH and 30CH potencies, which led the researchers to suggest new experiments using MB-4 rock flour in other potencies and crops.

3 MATERIAL AND METHODS

The experiment was carried out at the Agricultural Experimental Unit of the Federal University of Acre (UFAC), Rio Branco Campus, located in the city of Rio Branco-Acre, lasting 50 days from November 10 to December 29, 2014, marking the inauguration of this area as the first experiment to be set up.

3.1 OBTAINING SEEDS

We used pelleted seeds of two cultivars, "Verònica" and "Regina", from SAKATA (lot 90114, germination 95% and physical purity 99.9%) and TENOCSEED (lot TEA 1325-1A/3A, germination 95% and physical purity 99%).

3.2 SEEDLING PRODUCTION

The seedlings were grown in Styrofoam trays (23 for each cultivar) with 128 cells and sown using commercial substrate (Subras®) and one seed per cell (Figure 2).

After sowing, the trays were taken to the greenhouse and placed on a smooth wire hanging support made by the agricultural unit's technicians and assistants (Figure 3). During the seedling growth phase, the trays were irrigated daily by hand as required.

Figure 2: Sowing in Styrofoam trays (Source: personal archive).

Figure 3. Arrangement of the trays on the support (Source: personal archive).

3.3 OBTAINING TREATMENTS

The treatments were applied individually, consisting of the homeopathic medicines *Arnica montana, Calcarea carbonica, Carbo vegetabilis, Silicea terra, Phosphorus, Pulsatilla nigricans,* both purchased from Farmàcia Homeopàtica, in the Hahnemanian centesimal strengths 6CH, 12CH and 30CH and a homeopathic preparation in the same strengths as the other treatments (as described in the process of obtaining it), as well as two witnesses, 30% alcohol (ROSSI et al., 2007), which is the homeopathic carrier, and water.

3.3.1 Mother tincture and dynamizations of the homeopathic preparation Açai

This procedure was carried out to obtain the mother tincture, as there is no way to buy it in a pharmacy.

To obtain the açai solution, the mother tincture (MT) was first prepared and then the 6CH, 12CH and 30CH dynamizations. 100g of açai residue (leftover peel and seeds) was weighed and ground in a blender with 1000 mL of distilled water for approximately 1 minute. This mixture was then added to an amber glass bottle and stirred for 15 days, according to the method adapted from Mapeli (2010).

After this period, the mixture was filtered and then stored in an amber bottle for 15 days for further dynamization.

The 6CH, 12CH and 30CH dynamizations were made following the rules of the Brazilian Homeopathic Pharmacopoeia (1997). 30 mL amber glass flasks were used, in which 19.8

mL of the volume was filled with a 30% alcohol solution and 0.2 mL of the TM was used to make 100 succussions, obtaining açai 1CH. For açai 2CH, 0.2 mL of the 1CH was removed and added to a glass with 19.8 mL of 30% alcohol and shaken 100 times. The process was repeated until 30CH was obtained (Figures 4 and 5).

Figura 4. Dynamizations of the Açai preparation (Source: personal archive).

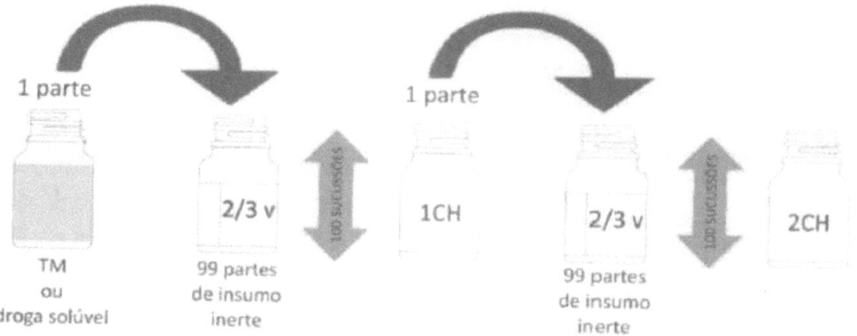

Figura 5. Schematic representation of the Hahnemannian method on a centesimal scale (Source: Carneiro et al, 2011a).

3.4 TREATMENT APPLICATIONS

To apply the treatments, the methodology described by Rossi (2005) was followed. Applications were made from the seventh day after the seedlings emerged (Figure 6), three times a week (on Mondays, Wednesdays and Fridays) until the twenty-eighth day (Figure 7), always at dusk, as described by Tichavsky (2007) in the Agrohomeopathy Manual.

Figura 6. Beginning of the treatments (Source: personal archive).

Figura 7. End of treatment (Source: personal archive).

Each tray received a single treatment. To prepare the irrigation solutions, 0.5 mL of each treatment was diluted in 500 mL of water (BONATO et al., 2014) and applied via spraying to the point of run-off.

The double-blind procedure indicated in the homeopathic experimentation protocol was adopted. The treatments were coded, remaining incognito to the applicator and evaluator and known only to the research administrator (CARNEIRO et al., 2011a).

3.5 SEEDLING DEVELOPMENT ANALYSIS

At the end of the treatments, development was assessed using the following variables: number of leaves (NF), height (ALT), length of the root system (CR), dry mass of the aerial part (MSPA), dry mass of the root system (MSR), neck diameter (DC), Dickson Quality Index (IQD) and number of plants developed in the field 15 days after transplanting.

Five seedlings were sampled from each repetition (Figure 8) and taken to the Plant

Production Laboratory. The number of leaves (NF) was counted and the diameter of the neck (DC) was measured using a digital caliper.

Figura 8. Seedlings collected for analysis (Source: personal archive).

The height of the seedlings (ALT) and the length of the root system (CSR) were obtained using a ruler graduated in millimeters. They were washed under running water to remove the substrate (Figure 9). The seedlings were then placed in labeled paper bags and dried in an oven with forced air circulation at 65°C (Figure 10). When the constant weight was reached, the dry mass of the aerial part (MSPA) and the dry mass of the roots (MSR) were determined on an analytical balance (Figure 11), following the method adapted from Rossi (2005).

Figure 9. Seedlings after washing (Source: personal archive).

Figure 10. Samples in the oven (Source: personal archive).

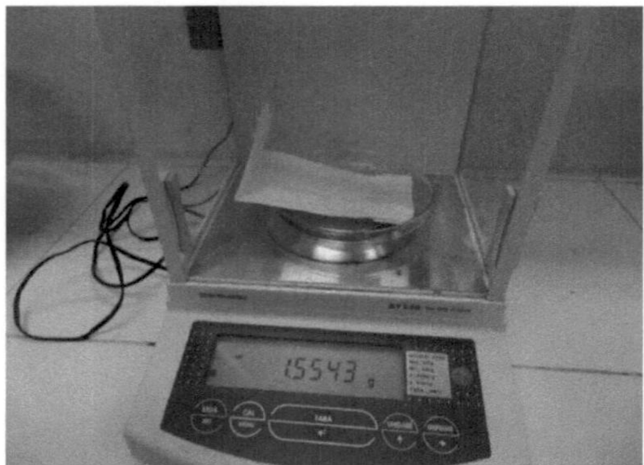

Figure 11. Weighing (Photo: personal archive).

To obtain the IQD, the methodology proposed by Freitas et al. (2013) was used, considering the indicators of dry mass of the aerial part, roots and total dry mass, height and neck diameter of the seedlings, according to equation **(1)**:

$$IQD = \frac{MST(g)}{\dfrac{H(cm)}{DC(cm)} + \dfrac{PMSPA(g)}{PMSRA(g)}}$$

Where: IQD = Dickson's development index, MST = total dry mass (g), H = height (cm), DC = neck diameter (cm), PMSPA = weight of aerial dry matter (g) and PMSRA = weight of root dry matter (g).

20

The remaining seedlings were transplanted into greenhouse beds filled with gravel and topsoil, spaced 30 cm apart **(Figure 12)**. After 15 days, the survival of the transplanted seedlings was assessed and the number of developed seedlings was determined (**Figure 13)**.

Figura 12. Transplanting the remaining seedlings (Photo: personal archive).

Figura 13. Seedling survival analysis (Photo: personal archive).

3.6 STATISTICAL ANALYSES

The statistical analysis used an inter-randomized design, with 21 treatments and two controls for each cultivar and four replications. Each plot consisted of a quarter of a 128-cell Styrofoam tray. The means were compared using the Scott-Knott test at a 5% probability level using the SISVAR statistical program (FERREIRA, 2000).

4 Results and discussion

The treatments tested in the experiment altered the organizational pattern of the lettuce seedlings, and the response was evaluated by the change in the development variables analyzed.

4.1 Cultivar 'Verònica'

In the development analysis, the height variable was positively influenced in relation to the controls by the medicines *Arnica montana* (6CH and 30CH), *Calcarea carbonica* (12CH), *Carbo vegetabilis (6CH)*, *Silicea terra (6CH* and 12CH), *Phosphorus (12CH)*, *Pulsatilla nigricans (12CH)* and the homeopathic preparation of açai (12CH) (**Table 1**).

Specifically in relation to the Arnica treatment, the similar results of the extreme potencies 6CH and 30CH support Bonato's (2004) assertion that the use of homeopathics on plants has a wave-like response effect. For this reason, it is essential when experimenting with plant homeopathy to work with several potencies of the same medicine, as a single dilution loses the information on the wave response (GONÇALVES et al., 2010).

Rossi and colleagues (2006) tested different potencies of the *Carbo vegetabilis* medicine on lettuce seedlings grown in two different environments (full sun and shade) and observed that the 6CH and 200CH potencies increased the height of plants grown in the sun. These studies confirm the wave effect provided by the potencies of the same homeopathy on plant growth.

However, Carvalho et al. (2005), when working with artemisia plants treated with *Arnica montana* in different potencies, did not obtain satisfactory results for the ALT parameter, as did Grisa et al. (2007a), who applied Arnica in different potencies to lettuce plants.

TABLE 1. Height (ALT), Root system length (CR) in centimeters (cm), Number of leaves (NF) and Collar diameter (CR) in millimeters (mm) of 'Veronica' cultivar lettuce seedlings.

TREATMENTS	ALT (cm)	CR (cm)	NF	DC (mm)
Arnica montana 6CH	3,46 a	8,27 b	4,00 b	1,87 a
Arnica montana 12CH	3,12 b	7,95 c	4,40 a	1,39 c
Arnica montana 30CH	3,51 a	7,39 c	4,55 a	1,27 c
Calcarea carbonica 6CH	3,19 b	9,06 a	4,15 b	1,81 a

Calcarea carbonica 12CH	3,66 a	7,33 c	4,60 a	1,44 b
Calcarea carbonica 30CH	3,21 b	7,48 c	4,35 a	1,26 c
Carbo vegetabilis 6CH	3,43 a	9,65 a	4,40 a	1,84 a
Carbo vegetabilis 12CH	3,22 b	7,44 c	4,45 a	1,47 b
Carbo vegetabilis 30CH	2,89 b	7,06 c	4,00 b	1,17 c
Silicea terra 6CH	3,41 a	9,17 a	4,05 b	1,66 b
Silicea terra 12CH	3,57 a	7,49 c	4,55 a	1,26 c
Silicea terra 30CH	2,96 b	8,11 b	4,05 b	1,60 b
Phosphorus 6CH	2,92 b	8,16 b	4,25 b	1,35 c
Phosphorus 12CH	3,61 a	7,82 c	4,50 a	1,49 b
Phosphorus 30CH	2,87 b	7,13 c	4,10 b	1,06 d
Pulsatilla nigricans 6CH	3,14 b	7,94 c	4,35 a	1,17 c
Pulsatilla nigricans 12CH	3,61 a	7,43 c	4,60 a	1,48 b
Pulsatilla nigricans 30CH	2,89 b	6,48 c	4,20 b	1,10 d
Açai 6CH	2,94 b	7,39 c	4,10 b	1,07 d
Açai 12CH	4,01 a	7,23 c	4,60 a	1,50 b
Açai 30CH	3,02 b	8,39 b	4,00 b	1,64 b
Witness 1- alcohol	2,47 b	6,93 c	4,00 b	0,87 d
Witness 2- water	3,18 b	7,72 c	4,15 b	1,66 b
Average	3,23	7,78	4,27	1,41
CV (%)	9,45	7,04	6,34	10,72

Means followed by the same letter in the column do not differ by the Scott-Knott test, considering a nominal value of 5% significance.

It can be seen that among the treatments that had a positive influence on the height of the seedlings, açai 12CH stood out, as it had the best value (**Table 1**). In order to study the potential of local resources in the preparation of homeopathics, it was proposed to include this preparation. Andrade and Casali (2011) state that experimenting with new homeopathic preparations produced from local resources is of great value as it acts as a sustainability strategy, favoring the independence of producers.

Since decomposed açai residues are used as organic fertilizers because they are rich in carbon (TEIXEIRA et al., 2004), it can be inferred that the increase in ALT in the seedlings treated with the homeopathic preparation at 12CH occurred because they were offered the information of organic matter, via açai 12CH. This phenomenon is explained by Capra (1983) cited by Andrade et al. (2001), who state that matter and energy are the same and interconvertible, with only the frequency of vibration changing.

One of the worst performances in the ALT parameter was shown by the 30% alcohol control (**Table 1**). This shows that the signals obtained with the application of the drugs are not influenced by the use of the vehicle. Alcohol is responsible for cell lysis, even at a low concentration (30%) the information of intoxication remained and affected the growth of the seedlings. A similar result was found by Moraes (2009) when researching the growth and quality of clonal eucalyptus seedlings using homeopathic preparations.   the other treatments when they received the 6CH alcohol application.

In the evaluation of the CR variable, *Calcarea carbonica, Carbo vegetabilis and Silicea terra,* all in the 6CH potency, showed a statistically significant difference from the two witnesses (**Table 1**). In this case, the hypothesis of an increase in CR should be considered, as the Calcarea, Carbo and Silicea molecules are still present in the drug at this potency. It is only from the 12CH potency onwards that Avogadro's constant is exceeded, i.e. homeopathies are probabilistically devoid of molecules, and the hypothesis of a dynamic physical effect, according to Casali et al. (2006), applies.

Studies related to the length of the root system were demonstrated by Bonfim et al. (2008) when they checked the influence of different dynamics of *Arnica montana* on the rooting of rosemary and *Lippia alba* (lemon balm), they concluded that the 6CH potency increased the RC of rosemary and lemon balm. Hamman et al. (2003), when studying the effects of dynamized gibberellic acid on barley germination using 3 batches with 3 levels of vigor, observed that the batch with medium vigor increased the length of the roots after applying the homeopathy. It is well known that seedlings with more developed roots can withstand transplanting better than those with more robust aerial parts (KARCHI et al., 1992).

With regard to NF, the treatments Arnica (12CH and 30CH), Calcarea (12CH and 30CH), Carbo (6CH and 12CH), Silicea (12CH), *Phosphorus (12CH)*, Pulsatilla (6CH and 12CH) and açai (12CH) stood out, showing a greater number of leaves when compared to the other treatments and the control (**Table 1**).

Similarly, Grisa et al. (2007a) tested the homeopathic medicine *Arnica montana* in 12CH and 30CH potencies on lettuce plants, but found no significant effects on NF. This contradiction probably reflects the different sowing conditions and climate of the experiments.

Rolim et al. (2002) observed that *Silicea terra* 30CH promoted a 60% increase in the number of leaves in passion fruit. This disagrees with the results found in this experiment, where only Silicea 12CH showed significant results (**Table 1**). A satisfactory response for increasing the number of leaves was also reported by Datta (2006) when checking the effects of Cina in combating *Meloidgyne incognita* in mulberry trees. It is common to see the same medicine causing different effects, depending on the strength and species tested, a case that can also be observed in this work. This non-linear effect of homeopathic science is also reported by Bonato and Peres (2007).

For the DC variable, the treatments *Arnica montana, Calacarea carbonica* and *Carbo vegetabilis,* all in the 6CH potency, showed statistically superior results to the other treatments and the two witnesses (**Table 1**). The treatments Calcarea 12CH, Carbo 12CH, Silicea 6CH and 30CH, *Phosphorus 12CH and* Pulsatilla 12CH and açai 12CH and 30CH showed similar results to each other and to the water control, differing only from the 30% alcohol control. Statistically inferior results were observed in the seedlings that received the medicines *Phosphorus* and *Pulsatilla nigricans* at 30CH, the açai preparation at 6CH and the 30% alcohol control. The other treatments differed only from the alcohol control.

Bonato and Silva (2003), working with radishes, investigated the effects of *Sulphur* and observed that the average diameter of the roots that received the 12CH, 30CH and 1MCH treatments was three times greater than the diameter of the roots of the control plants.

Taiz and Zeiger (2004) state that seedlings with a larger neck diameter are more likely to survive after transplanting, as they have a greater capacity to form and grow new roots. Therefore, CD is a good indicator of development.

In the evaluation of the MSPA parameter, the treatments *Arnica montana* 6CH, 12CH and 30CH; *Pulsatilla nigricans* 6CH and 12CH; *Calcarea carbonica, Carbo vegetabilis, Silicea terra, Phosphorus* and açai all at 12CH were the same as the water control, differing only from the alcohol control (**Table 2**). The others were the same as the alcohol control.

The increase in MSPA caused by the drug *Arnica montana* (6CH and 12CH) was also observed by Grisa et al. Rossi et al. (2003), when applying Carbo 30CH to lettuce, observed that applications every 48 hours increased leaf weight. In another study, Rossi et al. (2006) applied the same medicine, but in potencies of 6CH, 12CH, 30CH, 100CH and 200CH, also on lettuce seedlings grown in two different environments and found that the treatment in potency 100CH increased the dry mass of the aerial part of the seedlings grown in a shaded environment.

TABLE 2 - Aerial part dry mass (ASM), root system dry mass (RSM) in grams (g) and Dickson quality index (DQI) of lettuce seedlings cultivar Verònica'.

TREATMENTS	MSPA (g)	MSR (g)	IQD
Arnica montana 6CH	0,42 a	0,16 b	0,027 a
Arnica montana 12CH	0,46 a	0,24 a	0,029 a
Arnica montana 30CH	0,45 a	0,25 a	0,024 a
Calcarea carbonica 6CH	0,36 b	0,22 a	0,030 a
Calcarea carbonica 12CH	0,53 a	0,25 a	0,028 a
Calcarea carbonica 30CH	0,37 b	0,22 a	0,022 b
Carbo vegetabilis 6CH	0,39 b	0,22 a	0,030 a
Carbo vegetabilis 12CH	0,41 a	0,22 a	0,027 a
Carbo vegetabilis 30CH	0,32 b	0,19 b	0,019 b
Silicea terra 6CH	0,40 b	0,22 a	0,028 a
Silicea terra 12CH	0,49 a	0,26 a	0,025 a
Silicea terra 30CH	0,30 b	0,17 b	0,023 b
Phosphorus 6CH	0,39 b	0,23 a	0,027 a
Phosphorus 12CH	0,48 a	0,24 a	0,027 a
Phosphorus 30CH	0,31 b	0,17 b	0,017 b
Pulsatilla nigricans 6CH	0,42 a	0,23 a	0,023 b
Pulsatilla nigricans 12CH	0,47 a	0,25 a	0,027 a
Pulsatilla nigricans 30CH	0,36 b	0,21 a	0,021 b
Açai 6CH	0,32 b	0,21 a	0,018 b
Açai 12CH	0,56 a	0,28 a	0,029 a
Açai 30CH	0,26 b	0,18 b	0,022 b

Witness 1- alcohol	0,26 b	0,13 b	0,013 b
Witness 2- water	0,36 a	0,21 a	0,027 a
Average	0,39	0,22	0,024
CV (%)	21,21	13,70	17,57

In this experiment, the drug *Carbo vegetabilis*, only in the 12CH potency, showed satisfactory results, differing statistically from the 30% alcohol control, and when compared to the water control it showed a better average, but with no statistically significant difference. This contrast probably reflects the different sowing conditions, climate and frequency of application, since the same cultivar was used.

When evaluating the MSR variable, all the treatments were the same as the water control, except for *Arnica montana* 6CH and *Carbo vegetabilis, Silicea terra, Phosphorus* and açai, both at 30CH, which were similar to the alcohol control (**Table 2**). Satisfactory results for this variable at 12CH were demonstrated by Grisa et al (2007b) on beet plants submitted to the application of the homeopathic medicine *Staphysagria*.

According to Filgueira (2003), tissues rich in dry mass (DM) are favorable for good rooting and the resumption of plant development after transplanting.

The treatments *Arnica montana* (6CH, 12CH and 30CH); *Calcarea carbonica, Carbo vegetabilis, Silicea terra* and *Phosphorus* (both at 6CH and 12CH); and *Pulsatilla nigricans* and açai (both at 12CH) provided the highest values for the QDI, but did not differ from the control water, only from 30% alcohol (**Table 2**). Lower values were presented by the other treatments, which did not differ from control 1.

According to Fonseca (2000), the Dickson index is an important indicator of seedling quality, as its formula takes into account the vigor and balance of the distribution of phytomass in seedlings, weighting several important morphological measures.

4.2 Cultivar 'Regina'

The heights of the seedlings varied depending on the drugs and strengths, sometimes increasing and sometimes decreasing (**Table 3**). The seedlings treated with Arnica (6CH and 12CH), Calcarea (6CH), Carbo (6CH) and Silicea (6CH, 12CH and 30CH) were statistically superior in height when compared to the other homeopathic treatments, but did not differ from the two control treatments.

TABLE 3 Height (ALT), Root System Length (CR) in centimeters (cm), Number of Leaves (NF) and Collar Diameter (CR) in millimeters (mm) of lettuce seedlings cultivar 'Regina'.

TREATMENTS	ALT (cm)	CR (cm)	NF	DC (mm)
Arnica montana 6CH	3,81 a	8,22 b	7,55 c	2,15 c
Arnica montana 12CH	3,67 a	8,92 b	8,15 b	2,42 b
Arnica montana 30CH	3,11 c	8,64 b	8,15 b	2,28 c
Calcarea carbonica 6CH	3,76 a	8,09 b	7,70 c	2,23 c
Calcarea carbonica 12CH	3,45 b	8,13 b	8,10 b	2,5 b
Calcarea carbonica 30CH	3,28 b	8,67 b	7,95 b	2,37 c
Carbo vegetabilis 6CH	3,86 a	8,46 b	8,70 a	2,68 a
Carbo vegetabilis 12CH	3,44 b	8,78 b	7,90 b	2,45 b
Carbo vegetabilis 30CH	3,29 b	9,40 a	8,15 b	2,30 c
Silicea terra 6CH	3,88 a	8,65 b	9,15 a	2,67 a
Silicea terra 12CH	3,61 a	7,77 b	8,40 b	2,55 b
Silicea terra 30CH	2,70 a	8,82 b	7,35 c	1,94 c
Phosphorus 6CH	3,19 b	8,11 b	7,55 c	2,13 c
Phosphorus 12CH	3,32 b	9,08 a	7,80 b	2,27 c
Phosphorus 30CH	2,83 c	9,55 a	7,40 c	2,05 c
Pulsatilla nigricans 6CH	3,36 b	8,58 b	7,90 b	2,39 c
Pulsatilla nigricans 12CH	3,18 b	8,08 b	8,05 b	2,13 c
Pulsatilla nigricans 30CH	3,01 c	8,47 b	7,50 c	2,23 c
Açai 6CH	3,45 b	8,74 b	8,00 b	2,33 c
Açai 12CH	3,33 b	8,52 b	8,35 b	2,19 c
Açai 30CH	3,07 c	9,61 a	7,15 c	2,07 c
Witness 1- alcohol	3,77 a	8,07 b	9,05 a	2,57 b
Witness 2- water	4,10 a	9,81 a	8,80 a	2,90 a

Average	3,41	8,66	8,03	2,34
CV (%)	9,45	7,04	6,34	10,72

The drugs *Calcarea carbonica* and *Carbo vegetabilis* (both at 12CH and 30CH); *Phosphorus, Pulsatilla nigricans* and açai (both at 6CH and 12CH) produced seedlings with statistically similar heights. The treatments Arnica, *Phosphorus,* Pulsatilla and açai (both at 30CH) produced seedlings with statistically similar heights to each other and lower than the heights of the seedlings that received the other treatments.

Specifically in relation to the drug Calcarea at 6CH, Nunes (2013) when testing it on the development of *coriander* (*Coriandrum sativum* L.) cultivar "Verdâo", did not obtain satisfactory results, which does not exclude the possibility of testing other strengths or different crops, which may show positive effects.

Muller et al. (2009), stated that cases like this can be explained by the use of homeopathics in inappropriate strengths for the crop, since there are different effects for the same medicine when used in different concentrations.

In the rooting of rosemary and *Lippia alba,* Bonfim et al. (2008) demonstrated the influence of dynamizations of the drug *Arnica montana.* They obtained a significant increase in: Rosemary, in the 3CH and 6CH strengths in root length, and 6CH in the percentage and quality of rooting; Lippia, in the 3CH, 6CH and 12CH strengths in the number of branches, root length, and quality of rooting, and 6CH in the percentage of rooting.

In the physiology of Artemisia (*Tanacetum parthenium*), Carvalho et al. (2003) used *Arnica montana* and obtained an increase in fresh mass with potency 1D (1:10 dilution) and a reduction in parthenolide content with potencies 1D, 2D, 4D and 5D. In 2005, they observed a reduction in the parthenolide content per plant in the 3CH and 5CH potencies of Arnica (CARVALHO et al., 2005).

Tichavsky (2007) explains that homeopathic potencies are classified into 3 levels: low potencies up to 6CH; medium potencies from 6CH to 23CH and high potencies above 23CH, each of which can cause different effects. As a general rule, low strengths are indicated for solving acute physical problems, so the same homeopathy applied in medium potency can have a different response, as it is recommended for solving problems related to plant functions. High potencies, on the other hand, can have dramatic results, which is why they should be applied in less frequent doses (usually just once), because their effect will be long-lasting.

With regard to RC, Carbo and açai (both at 30CH) and *Phosphorus* (12CH and 30CH), showed the best lengths, and differed statistically from the 30% alcohol control. The other treatments produced seedlings with root system lengths that were similar to each other and to control 1 (**Table 3**).

Marques et al. (2008) found that increasing strengths of the homeopathic preparation of citronela (*Cymbopogon winterianus*) interfered with the growth and germination of *Sida rhombifolia* (a medicinal plant popularly known as guanxuma or vassourinha) and found that all the strengths used stimulated root growth, including 30CH. The same behavior was observed in this trial, in which among the homeopathies, the 30CH dynamization of the homeopathic preparation of açai was the one that provided the greatest RC in the seedlings. In this way we can confirm the action of non-molecular preparations (above 12CH the Avograd number is extrapolated) on plants, as well as the effect of homeopathic preparations.

Matter is a form of condensed energy. Dynamization is a way of releasing this energy, leaving the original information of the matter in the homeopathic solvent. This information is more important than the molecule (active ingredient) itself, and is stored by the vehicle (solvent) of the homeopathic medicine (CASALI et al., 2006; LISBOA et al., 2005; CAMPOS, 1994).

Endler et al. (1994) also explains that even if the original molecule is not present to promote a biological effect, the biomolecular information can be transmitted through the solvent. Studies demonstrating the effects of high dynamics on plants were carried out by Rossi (2005) in organic strawberry cultivation, where he observed that Carbo at 30CH increased seedling production. And by Deboni et al. (2008) on the germination of seeds of two black bean cultivars, where the authors observed an increase in the emergence of seedlings with the *Arnica montana* 30CH treatment.

In the NF parameter, *Carbo vegetabilis* and *Silicea terra,* both in the 6CH potency, showed the best results among the treatments, but were not statistically different from the 30% alcohol and water controls (**Table 3**).

In contrast to Rolim et al. (2002), who obtained a 60% increase in the number of leaves in passion fruit with the application of the *Silicea terra* 30CH medicine, in the present study a reduction in NF was observed in the seedlings that received the 30CH dynamization of this medicine. As mentioned above, the effect of the same drug can vary according to its potency, culture and growing environment.

Among the results obtained with the use of *Silicea terra* is the favoring of normal development, in cabbage (*Brassica oleracea* L.) it promoted the growth of numerous, long roots and a greater number of vigorous leaves and in cacao (*Theobroma cacao* L.) it favored germination (REZENDE, 2009).

Similar effects to those found in the 'Verònica' cultivar were observed in the 'Regina' cultivar, which also showed superior CD results with the 6CH potency. The Carbo and Silicea (6CH) treatments performed well and differed only from the alcohol-free control (**Table 3**).

Muller et al. (2009), who examined the effects of homeopathic preparations on the development of the picâo-preto (principle of equals), found that the 30CH strength promoted an increase in stem diameter. This effect was not found for the DC variable in any of the treatments applied in this trial.

It was observed that the *Phosphorus*, Pulsatilla and açai treatments showed no variation in response to any of the three strengths applied. Research using other strengths could indicate contrary responses.

In relation to the drug *Phosphorus*, Moraes (2009), when studying the production of forest seedlings, observed that 12CH dynamization increased the quality standard of eucalyptus seedlings based on the characteristics desirable for the field.

As for *Pulsatilla nigricans*, the studies carried out by Muller and Toledo (2013) on tomatoes showed an increase in the number of fruits per plant that received this drug.

Phosphorus, Pulsatilla and açai (all at 6CH, 12CH and 30CH), *Arnica montana and Calcarea carbonica* (6CH and 30CH), *Carbo vegetabillis* and *Silicea terra* (30CH), showed statistically inferior results for CD when compared to the two control and other treatments (**Table 3**).

In the evaluation of MSPA, although the treatments Arnica (6CH), Calcarea (12CH), Carbo and Silicea (both at 6CH and 12CH) did not show any statistically significant variation from the control group, they did differ from the others (**Table 4**).

In work carried out by Pulido et al. (2014), the drug *Silicea terra had a* similar effect. They observed that this homeopathy led to an increase in the dry matter of cabbage heads.

In terms of root system dry mass, the treatments Carbo (6CH) and *Silicea terra* (6CH and 12CH) showed the best values for RSM, differing statistically only from the 30% alcohol control (**Table 4**). Toledo (2009), working with tomato plants, obtained satisfactory effects for the increase in RMS with the 6CH and 12CH potencies, but in this trial he used the *Sulphur* drug.

Higher values for the IQD of the seedlings were observed with the application of the treatments *Carbo vegetabilis* and *Silicea terra* at 6CH, which showed a significant difference when compared to the alcohol control, but did not differ from the water control (**Table 4**). In plants, both Carbo and Silicea are useful medicines for making plants stronger and more robust, i.e. a general tonic (CARNEIRO et al., 2011a; BONATO et al., 2014). One possible explanation for these satisfactory results with 6CH Carbo and Silicea is the chemical effect of the substances present in this dynamization.

As mentioned above, the IQD is considered a good indicator of seedling quality, as it takes into account the robustness and balance of the distribution of the seedlings' phytomass, weighting several important parameters.

TABLE 4 - Dry mass of the aerial part (MSPA), dry mass of the root system (MSR) in grams (g) and Dickson's quality index (IQD) of lettuce seedlings cultivar 'Regina'.

TREATMENTS	MSPA (g)	MSR (g)	IQD
Arnica montana 6CH	0,50 a	0,34 b	0,044 c
Arnica montana 12CH	0,42 b	0,34 b	0,046 c
Arnica montana 30CH	0,43 b	0,29 c	0,048 c
Calcarea carbonica 6CH	0,44 b	0,32 b	0,041 c
Calcarea carbonica 12CH	0,46 a	0,31 b	0,050 b
Calcarea carbonica 30CH	0,43 b	0,33 b	0,050 b
Carbo vegetabilis 6CH	0,55 a	0,37 a	0,058 a
Carbo vegetabilis 12CH	0,47 a	0,28 c	0,048 c
Carbo vegetabilis 30CH	0,39 b	0,32 b	0,046 c
Silicea terra 6CH	0,61 a	0,40 a	0,063 a
Silicea terra 12CH	0,48 a	0,35 a	0,054 b
Silicea terra 30CH	0,30 b	0,25 c	0,037 d
Phosphorus 6CH	0,33 b	0,23 c	0,035 d

Phosphorus 12CH	0,37 b	0,30 b	0,043 c
Phosphorus 30CH	0,28 b	0,25 c	0,036 d
Pulsatilla nigricans 6CH	0,41 b	0,29 c	0,045 c
Pulsatilla nigricans 12CH	0,39 b	0,31 b	0,043 c
Pulsatilla nigricans 30CH	0,36 b	0,26 c	0,042 c
Açai 6CH	0,41 b	0,31 b	0,045 c
Açai 12CH	0,45 b	0,31 b	0,045 c
Açai 30CH	0,35 b	0,26 c	0,038 d
Witness 1- alcohol	0,53 a	0,33 b	0,053 b
Witness 2- water	0,53 a	0,38 a	0,059 a
Average	0,43	0,31	0,046
CV (%)	21,21	13,70	17,57

Means followed by the same letter in the column do not differ by the Scott-Knott test, considering a nominal value of 5% significance.

4.3 SURVIVAL OF SEEDLINGS AFTER TRANSPLANTING

15 days after transplanting into the beds, 100% of the seedlings ('Verònica' and 'Regina') had survived. It is possible that this survival rate is related to the balance provided by the application of homeopathic medicines and preparations. To prove this, it is necessary to monitor the development of these plants in the field, checking their behavior under adverse environmental conditions, such as high temperatures, excessive rainfall, diseases and pest attacks (Rossi, 2005).

CONCLUSION

In the analysis of the development of the 'Verònica' cultivar, it was observed that in almost all the parameters evaluated, the medicine *Carbo vegetabillis* 6CH, followed by *Calcarea carbonica* 12CH, *Silicea terra 12CH, Phosphorus 12CH, Pulsatilla nigricans* 12CH and the preparation of açai 12CH helped the seedlings to perform better. The medicines *Silicea terra* 6CH and *Carbo vegetabillis* 6CH influenced all the variables studied in the 'Regina' cultivar, except for the growth of the root system.

FINAL CONSIDERATIONS

Non-inearity in homeopathic crops seems to be common. These reports are in line with the results observed in this study, highlighting the different effects that each dynamization of the same medicine can have.

As the work with homeopathic crops is pioneering, for the non-significant results observed in the development variables, it is recommended that other strengths, other frequencies and forms of application be analyzed, as well as considering the pharmacopoeia used, as the results may differ.

It is believed that homeopathies interfere with plant metabolism, demonstrating the need and importance of further research into this alternative therapy for agricultural crops.

BIBLIOGRAPHICAL REFERENCES

BRAZILIAN ASSOCIATION OF HOMEOPATHIC PHARMACISTS. **Manual of technical standards for homeopathic pharmacy:** expansion of the technical and practical aspects of homeopathic preparations. 4. Ed. Curitiba: ABFH, 2007.182 p.

ALVES, A. S. **Homeopathy *Sepia, Calcarea carbonica* and homeopathic preparation Thiourea, in the development of reproducers and tadpoles of bullfrog (*Lithobates catesbeianus*).** 2014. 76 f. Dissertation (Master's Degree in Animal Biology) - Federal University of Viçosa, Viçosa, 2014.

ANDRADE, F.M.C. et al. Growth and coumarin production in chambà plants (*Justicia pectoralis* Jacq.) treated with isotherapic. **Rev. Bras. Plantas Medicinais**, Botucatu, v.14, especial, p.154-158, 2012.

ANDRADE, F. M. C.; CASALI, V. W. D. Homeopathy, agroecology and sustainability. **Revista Brasileira de Agroecologia**, Porto Alegre, v. 6, n. 1, p. 49-56, 2011.

ANDRADE, F. M. C. Tecnologia e Aplicaçâo da Homeopatia na Horticultura. 2010. **Revista Horticultura Brasileira**, Brasilia, DF, v.28, n. 2, p.85-91,2010.

ANDRADE, F. M. C. Strategies and methods for implementing homeopathy on rural properties. In: SEMINARIO SOBRE CIÊNCIAS BASICAS EM HOMEOPATIA, 8., 2007, Lages. **Proceedings...** Lages: EPAGRI, 2007. p. 27-32.

ANDRADE, F.M.C. et al. Effect of homeopathies on the growth and production of coumarin in chambà (*Justicia pectoralis* Jacq.). **Revista Brasileira de Plantas Medicinais**, v.4, p.19-27, 2001.

ANDRADE, F.M.C. **Homeopathy on the growth and production of coumarin in shrub *Justicia pectoralis Jacq*.** 2000. 214p. Dissertation (Master's Degree - Area of Concentration in Plant Science) - Department of Plant Science, Federal University of Viçosa, Viçosa, 2000.

ARAÙJO, M. O. et al. Effect of doses of homeopathic preparations of moringa (*Moringa oleifera*) in water. In: BRAZILIAN CONGRESS OF AGROECOLOGY, 8, 2013, Porto Alegre. **Proceedings...** Porto Alegre: Cadernos de Agroecologia, v. 8, n. 2, Nov. 2013. p. 1-4.

BOFF, P.; GIESEL, A. Plant homeopathy and the management of leaf-cutting ants. In: BOFF, P. (coord.) **Healthy agriculture: from disease, pest and parasite prevention to non-residual therapy.** Lages: EPAGRI; UDESC, 2008. p. 51-56.

BONATO, C.M. et al. **Simple homeopathy: alternatives for family farming**. 4. ed. Marechal Càndido Rondon: Lider, 2014. 46p.

BONATO, C. M. Homeopathy in agriculture. In: ENCONTRO BRASILEIRO DE HOMEOPATIA NA AGRICULTURA, 1., 2009, Campo Grande. **Proceedings...** Campo Grande: AMVH, 2009.

BONATO, C. M. et al. Homeopathic drugs *Arsenicum album* and *Sulphur* affect the growth and essential oil content in mint (*Mentha arvensis* L.). **Acta Scientiarum Agronomy**, Maringà, v. 31, n.1, p. 101-105, 2009.

BONATO, C.M.; PERES, P.G.P. Homeopathy in vegetables. In: SEMINARIO SOBRE CIÊNCIAS BASICAS EM HOMEOPATIA, 7, 2007, Lages. **Proceedings...** Lages: CAV/UDESC; EPAGRI, 2007. p.41-59.

BONATO, C. M. Mechanism of action of homeopathy in plants. In: V SEMINARIO BRASILEIRO SOBRE A HOMEOPATIA NA AGROPECUARIA ORGÀNICA. 2003, Toledo, PR.Viçosa - FUNARBE - Universidade Federal de Viçosa, 2004, p. 17-44.

BONATO, C. M.; SILVA, E. P. Effect of the homeopathic solution Sulphur on the growth and productivity of radish. **Acta Scientiarum Agronomy**, Maringà, v. 25, n.2, p. 259-263, 2003.

BONFIM, F.P.G. et al. Use of homeopathic *Arnica montana* for the issuance off roots os *Rosmarinus officinalis* L. and *Lippia alba* (Mill) N.E.Br. **International Journal of High Dilution Research**, Guaratinguetà, v.7, n. 23, p. 113-117, 2008.

BRAZIL. **Normative Instruction No. 46** of October 8, 2011. Regulates organic animal and plant production. Diàrio da Repùblica Federativa do Brasil, Brasilia, DF, October 8, 2011 - Section I, p. 11 to 14.

BRAZIL. **Normative Instruction No. 07**, of May 17, 1999. Regulates the production of organic plant and animal products. Diàrio da Repùblica Federativa do Brasil, Brasilia, DF, May 17, 1999 - Section I, p. 11 to 14.

CAMPOS, J. M. **O eterno plantio:** reencontro da medicina com a natureza. Sâo Paulo - SP:Cultrix, 1994. 247 p.

CAPRA, R. S. **Effect of homeopathic preparations and growing environment on the production of flavonoids and saponins by carqueja plants**. 2011. 53 f. Dissertation (Master's Degree in Plant Science) - Federal University of Viçosa, Viçosa, 2011.

CARNEIRO, S. M. de T. P. G. et al. **Homeopathy principios e aplicações na**

Agroecologia. Londrina: IAPAR, 2011. 234p.

CARNEIRO, S. M. T. P. G. et al. Effect of homeopathic medicines, isotherapics and substances in high dilutions on plants: bibliographical review. **Revista de Homeopatia.** v.74, n.1/2, p. 9-32, 2011b.

CARVALHO, L.M.et al. Effect of *Arnica montana* homeopathy, in centesimal potencies, on artemisia plants. **Revista Brasileira de Plantas Medicinais,** Botucatu, v.7,n.3, p. 33-36, 2005.

CARVALHO, L.M.et al. Effect of Homeopathy on the recovery of Artemisia plants [*Tanacetum parthenium* (L.) Schultz-Bip] submitted to water deficiency. **Revista Brasileira de Plantas Medicinais,** Botucatu, v.6. n. 2, p. 20-27, 2004.

CARVALHO, L.M. et al. Effect of decimal potencies of *Arnica montana* homeopathy on Artemisia plants. **Revista Brasileira de Plantas Medicinais,** Botucatu, v.6,n.1, p. 46-50, 2003.

CASALI, V. W. et al. Homeopathy, agroecology and sustainability. **REVISTA BRASILEIRA DE AGROECOLOGIA,** Porto Alegre, v. 6, n. 1, p. 49-56, 2011.

CASALI,V.W.D. et al. **Homeopathy: bases and principles.** Viçosa, MG: Universidade Federal de Viçosa/DFT, 2006. 149 p.

CASALI V.W.D. Utilization of homeopathy in vegetables. In: SEMINARIO Brasileiro sobre homeopatia na agropecuaria orgànica, 5, 2004, Toledo. **Proceedings...** Toledo, p.89-117. 2004.

CoSTA, N. C. C. et al. Homeopathy: A fundamental therapeutic field in the veterinary care of production animals. **Revista Salus-Guarapuava (PR),** Guarapuava - PR, v.3, n. 2, p. 75-89, Jul./Dec. 2009.

DATTA, S. C. Effects of *Cina* on root-knot disease of mulberry. **Homeopahy,** London, V. 95, p. 98-102, 2006.

DEBONI, T. C.et al. **Action of homeopathy on bean germination.** Campinas: IAC, 2008 (Documentos, 85). 1 CD-RoM.

ENDLER p. C.et al. The effect of highly diluted agitated thyroxine on the climbing activity of frogs. **J. Vet. Hum. Tox.** n. 36, p. 56-59, 1994.

FARMACOPÉIA HOMEOPATICA BRASILEIRA, Sao Paulo: Andrei Editora, 1977. 115p.

FAZOLIN, M. et al. Use of homeopathic medicines to control *Cerotoma tingomariannus*

Bechyné (Coleoptera, Chrysomelidae) In: EMBRAPA, 2002, Rio Branco. Proceedings...Rio Branco: Acre, 2002.

FERREIRA, F. A. **SISVAR system for statistical analysis**. Lavras: Federal University of Lavras, 2000. Available at:

<http://www.dex.ufla.br/~danielff/softwares.htm.>. Accessed on: May 20, 2015.

FERREIRA, I. C. P. V. et al. Homeopathic preparations, barbatim extract and cow urine in the control of pineapple fusariosis. **Cadernos de Agroecologia**, v. 4, n. 1, 2009.

FILGUEIRA, F. A. R. **Novo manual de olericultura:** agrotecnologia moderna na produçâo e comercializâo de hortaliças. 2. ed. Viçosa: UFV, 2003.

FONSECA, E.P. **Quality standard of seedlings of *Trema micrantha* (L.) Blume., Cedrela fissilis Veli. and *Aspidosperma polyneuron* Müll Arg. produced under different shading periods**. Thesis (Doctorate in Agronomy) - Universidade Estadual Paulista, Jabotical, 2000.

FREITAS, G. A. et al. Lettuce seedling production as a function of different substrate combinations. **Revista Ciência Agronômica**, Fortaleza, v. 44, n. 1, p. 159-166, jan-mar. 2013.

GONÇALVES, P. A. S. et al. Homeopathic preparation of wormwood, *Artemisia vulgaris* L., in the management of thrips and its effect on onion production in an organic system. **Rev. Bras. de Agroecologia**, Porto Alegre, v. 5, n.2, p. 16-21, 2010.

GRISA, S. et al. Lettuce growth and productivity under different potencies of the homeopathic medicine *Arnica montana*. **Revista Brasileira de Agroecologia**, Porto Alegre, v.2, n.2, p.1050-1053, 2007a.

GRISA, S. et al. Quantitative analysis of beet plants treated with homeopathic preparations of *Staphysagria*. **Revista Brasileira de Agroecologia**, Porto Alegre, v.2, n.2, p.1046-1049, 2007b.

HAHNEMANN, S. **Organon da arte de cura**. 6th Ed. Sâo Paulo: Robe Editorial, 1996.

HAMMAN, B. et al. Homeopathically prepared gibberellic aid and barley seed germination. **Homeopathy,** London, v. 92, p. 140-144, 2003.

HENZ, G. P.; SUINAGA, F. **Types of lettuce grown in Brazil.** Brasilia: Embrapa Hortaliças, 2009. 7p. (Embrapa Hortaliças. Circular Técnica, 75).

JOSÉ, W. K.; CUÉLLER, J. O. O. Growth and productivity of lettuce (*Lactuca sativa* L.)

under different potencies of the homeopathic preparation of MB-4 rock flour. **Rev. Brasileira de Agroecologia,** v. 4, n.2, p. 4541-4544, nov. 2009.

KARCHI, Z. et al. Growth of containerized lettuce transplants supplemented with varying concentrations of nitrogen and phosphorus. **Acta Horticulturae**, v. 319, p. 367-370, 1992.

KHANNA, K.K.; CHANDRA, S. Control of tomato fruit by *Fusarium roseum* with homeopathic drugs. **Indian Phytopathology**, India, v.29, p.269-272, 1976.

KHUDA-BUKHSH, A. R. Laboratory research in homeopathy: pro. **Integr Cancer Ther**. v. 5, n.4, p.320-32, 2006.

LÉDO, F. J. S. et al. Performance of lettuce cultivars in the state of Acre. **Horticultura Brasileira**, Brasilia, v. 18, n. 3, p. 225-228, nov. 2000.

LISBOA, S. P. et al. **New vision of living organisms and balance through homeopathy.** Viçosa - MG, 2005. 103 p.

LONDRES, F. **Agrotóxicos no Brasil:** um guia para açâo em defesa da vida. Rio de Janeiro: AS-PTA, 2011. 190 p.

LUIS, S.; MORENO, N. **Efecto de cinco medicamentos homeopàticos em la producción de peso fresco, en Cebollin (*Allium fistolosum*).** 2007. Available at: http://www.comenius.edu.mx/CincomedicamentoshomeopaticonenCebollin.pdf

MARQUES, R.M.et al. Effect of high dilutions of Cimbopogon winteranus Jowitt (citronela) on the germination and growth of seedlings of Sida rhombifolia.

International Journal of High Dilution Research, Guaratinguetà, v. 7, n. 22, p. 3135, 2008.

MAPELI, N.C. et al. Repellency of Asciamonusteorseis (Latreille) (Lepidoptera: Pieridae) exposed to homeopathic solutions. **Revista Agrarian**, v.3, n.8, p. 119-125, 2010.

MORAES, L. C. A. **Growth and quality of clonal eucalyptus seedlings with the application of homeopathic preparations.** 2009. 54 f. Dissertation (Master's Degree in Plant Science) - Federal University of Viçosa, Viçosa, 2009.

MODOLON, T. A. **High dilution preparations for the phytosanitary and post-harvest management of tomato plants.** 2010. 78 f. Dissertation (Master's Degree in Plant Production) - Center for Agro-Veterinary Sciences, Santa Catarina State University, Lages, 2010.

MÜLLER, S. F.; TOLEDO, M. V. Homeopathy in tomato production in protected cultivation.

In: Brazilian Congress of Agroecology, 7. 2013, Porto Alegre. **Proceedings...** Porto Alegre: Cadernos de Agroecologia, v. 8, n. 2, nov. 2013. p. 1-4.

MÜLLER, S. F. et al. Effect of homeopathic preparations on the development of picâo-preto. **Revista Brasileira de Agroecologia**, Porto Alegre, v. 4, n. 2, p. 2533-2536, 2009. Available at:

<http://www6.ufrgs.br/seeragroecolgia/ojs/viewartcle.php?id=3467&layout=abstract>. Accessed on: June 28, 2015.

NUNES, F. J. Observation of the influence of 6CH *Calcarea carbonica* on the growth and development of *coriander* (*Coriandrum sativum* L.) cultivar "Verdâo". In: CONGRESSO BRASILEIRO DE AGROECOLOGIA, 8, Porto Alegre. **Proceedings...** Porto Alegre: Cadernos de Agroecologia, v. 8, n. 2, Nov. 2013. p. 1-5.

NUNES, R. DE O. **Tannin content in *Shagneticola triobata* (L.) Pruski with the application of *Sulphur* homeopathy**. 2005. 92 f. Thesis (Doctorate in Plant Science) - Federal University of Viçosa, Viçosa, 2005.

OLIVEIRA, H. J. et al. Radiostatic diagnosis and use of homeopathy in a Cupuaçu growing area. In: CONGRESSO BRASILEIRO DE AGROECOLOGIA, 7., 2011, Fortaleza. **Anais...**Cadernos de Agroecologia, v. 6, n. 2, p. 1-6, dec. 2011.

Available at: http://ainfo.cnptia.embrapa.br/digital/bitstream/item/53567/1/11997- 51872-1-PB.pdf. Accessed on: 6 Apr. 2015

OLIVEIRA, J. S. B. **Homeopathic medicines on the in vitro growth of *Pseudocercospora griseola* and on the physiology and biochemistry of bean plants**. 2012. 141 f. Dissertation (Master's Degree in Agronomy) - State University of Maringà, Maringà, 2012.

PULIDO, E. E. et al. High dilution preparations in organic cabbage production. **Horticultura Brasileira**, v. 32, n. 3, 2014.

REZENDE, J. M. (Org.) **Caderno de Homeopatia**. 3. ed. Viçosa: Universidade Federal de Viçosa / Departamento de Fitotecnia, 2009. 51p.

ROLIM, P. R. R. WORLD PANORAMA OF AGROHOMEOPATHY. In: Encontro Brasileiro de Homeopatia na Agricultura, 1. Campo Grande- MS.2009. Available at: http://www.cesaho.com.br/biblioteca_virtual/arquivos/arquivo_408_cesaho.pdf. Accessed on: May 25, 2015.

ROLIM, P. R. R. et al. Passion fruit crop management without the use of conventional agrochemicals. In: REUNIÂO TÉCNICA DE PESQUISA EM MARACUJAZEIRO, 3. **Anais...** Viçosa: UFV, p. 113, 2002.

ROSSI, F.; HAHNEMANN, C. F. S. Fundamentals of agrohomeopathy. In: I BRAZILIAN MEETING OF HOMEOPATHY IN AGRICULTURE. Campo Grande, Brazil, 2009.

ROSSI, F. et al. Organic potato cultivation with the application of homeopathic preparations. In: Brazilian Congress of Agroecology. 5., 2007. **Proceedings...** Porto Alegre: Rev. Brasileira de Agroecologia, v. 2, n. 2, Oct. 2007.

ROSSI, F.; MELO, P. C. T.; AMBROSANO, E. J.; GUIRADO, N.; SCHAMMASS, E.

A. The application of the homeopathic medicine *Carbo vegetabilis* and the development of lettuce seedlings. **Cultura Homeopâtica**, Sâo Paulo, v. 5, n. 17, 2006.

Available at: www.feg.unesp.br/~ojs/index.php/ijhdr/article/../173/177. Accessed on: March 25, 2015.

ROSSI, F. **Application of homeopathic preparations to strawberries and lettuce for agroecological cultivation**. 2005. 79 f. Dissertation (Master's Degree in Agronomy) - Luiz de Queiroz College of Agriculture, Piracicaba, 2005.

ROSSI, F.; et al. Basic experiences of homeopathy in plants. Contribution of plant research to the consolidation of homeopathic science. **Cultura Homeopâtica**, v.3, n.7, p. 12-13, 2004.

ROSSI, F. et al. Application of Carbo vegetabilis homeopathic solution and lettuce productivity. **Horticultura Brasileira**, Brasilia, v. 21, n. 2, supl., 2003. CD-ROM.

SANTOS, F. M. et al. Germination and growth of Brazilian lavender seedlings treated with *Phosphorus* homeopathy. In: CONGRESSO BRASILEIRO DE AGROECOLOGIA, 7, 2011, Fortaleza. **Anais...**Fortaleza: Cadernos de Agroecologia, v. 6, n. 2, Dec. 2011. p. 1-5.

SILVA, E. M. N. C. P. S. **Production and quality of organic lettuce grown with different soil preparations and shaded with passion fruit trellis, plastic and canvas, in Rio Branco-Acre**. 2010. 88 f. Dissertation (Master's Degree in Agronomy).

Federal University of Acre, Rio Branco, 2010.

TAIZ, L.; ZEIGER, E. **Fisiologia Vegetal**. Porto Alegre: Artmed Editora S/A, 2004. 438 p.

TEIXEIRA, L. B. et al. **Chemical characteristics of organic compost produced with organic waste, açai pits, grass and sawdust**. Belém: Embrapa Amazônia Oriental, 2004. 4p. (Embrapa Amazônia Oriental. Circular Técnica, 105).

TICHAVSKY, M. R. **Manual of Agrohomeopathy**. Monte Rei, 2007. Available at:<http://redeagroecologia.cnptia.embrapa.br/biblioteca/manejo/homeopatia/Manua l %20de%20Agrohomeopatia.pdf>. Accessed on: February 3, 2015.

TICHAVSKY, G. **Homeopathy: science and healing.** Sâo Paulo: Cultrix, 1980. 436 p.

TOLEDO, M. V. **Toxicity against *Alternaria solani,* control of black spot and on the growth of tomato plants (*Lycopersicum esculentum* Mill) by homeopathic medicines**. Master's degree dissertation. Marechal Càndido Rondon: State University of Western Paranà, 2009.

FEDERAL UNIVERSITY OF ACRE (UFAC). **17th Brazilian Seminar on Homeopathy in Organic Agriculture**. Rio Branco, 2015. Available at: http://www.ufac.br/portal/news/campus-floresta-sedia-evento-sobre-homeopatia-na-agropecuaria. Accessed on: 15/04/2015.

VITHOULKAS, G. **Homeopathy**: science and healing.Sâo Paulo: Editora Cultrix, 1980. 463p.

Printed by Books on Demand GmbH, Norderstedt / Germany